# Elementary Chemistry

## Its Properties & Its Changes

# Student Journal

Tom DeRosa
Carolyn Reeves

Elementary Chemistry

# MATTER

## Student's Journal

*Its Properties & Its Changes*

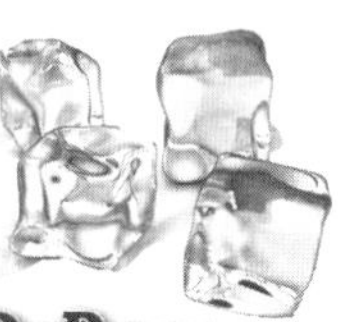

Tom DeRosa
Carolyn Reeves

First Printing: April 2009



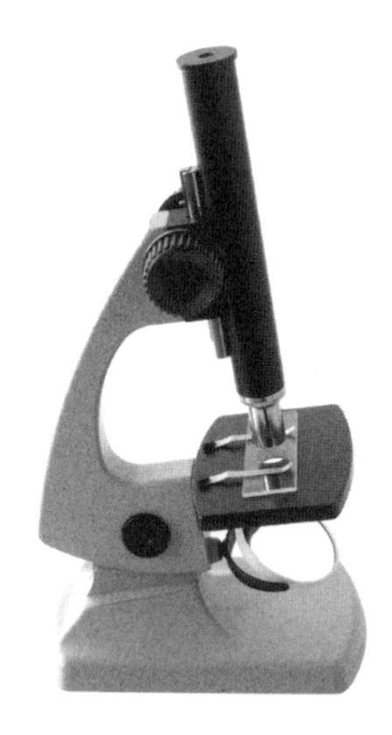

Master Books®
P.O. Box 726
Green Forest, AR 72638

Printed in the United States of America

Cover Design by Diana Bogardus and Terry White
Interior Design by Terry White

ISBN 10: 0-89051-559-X
ISBN 13: 978-0-89051-559-4
Library of Congress Control Number: 2009923589

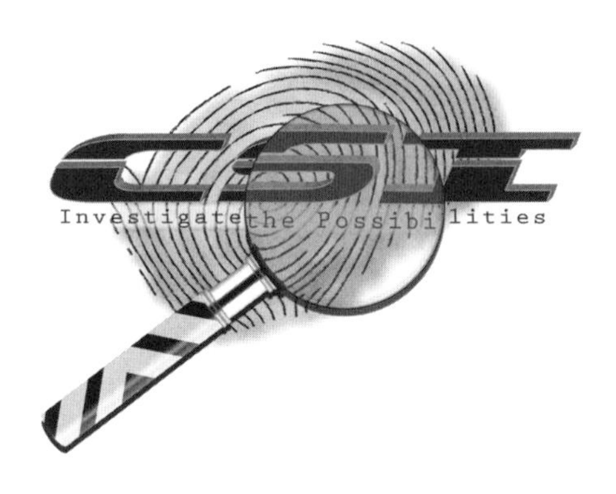


All Scripture references are New International Version unless otherwise noted.

All images from shutterstock.com

Please visit our website for other great titles: www.masterbooks.net

## Table of Contents



## Note to the Student

Record your ideas, questions, observations, and answers in the student book. Begin with "Thinking about This." After you read "Think about This," try to recall and note any experiences you have had related to the topic, or make notes of what you would like to learn.

Record all observations and data obtained from each activity.

You should do at least one Dig Deeper project each week. Your teacher will tell you how many projects you are required to do, but feel free to do more if you find an area that is especially interesting to you. The reason for the large number of projects is to give you choices. This allows you to dig deeper into those areas you are most interested in pursuing. Most of these projects will need to be turned in separately from the Student Answer Book, but uses the Student Answer Book to record the projects you choose to do along with a brief summary of each project and the date each is completed.

Record the answers to "What Did You Learn."

The Stumper's Corner is your time to ask the questions. Write two short answer questions related to each lesson that are hard enough to stump someone. Write your questions along with the correct answer or write two questions that you don't know and would like to know more about.

**These experiments require adult supervision. They have been specifically designed for educational purposes, with materials that are readily available. At their conclusion, please appropriately dispose of any by-products or food items included in the experiments.**

ACTIVITY 1

## Investigation #1
# The Physical Side of Chemicals

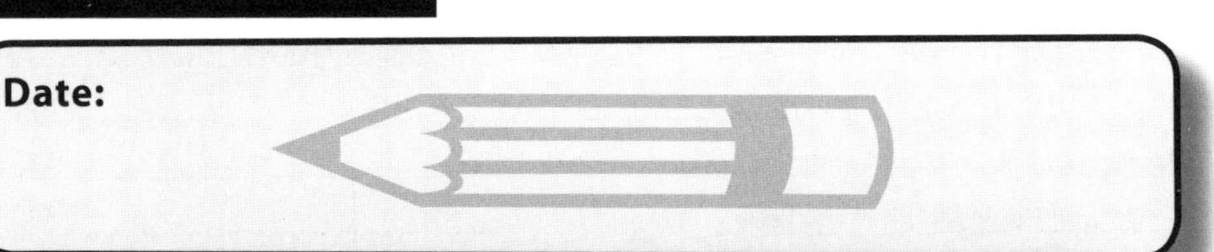

Date:

## The Activity: Procedure and Observations

Try to identify one of the items your teacher shows you on the basis of these physical properties: It is round. It is flat. You would not want to eat it. It would be hard to break. It is shiny. What is the item that has all of these properties?

Test each substance and complete the data table.

| Substance | Effect of a magnet | Float or sink in water | Soluble or insoluble in water | Color | Shiny or dull |
|---|---|---|---|---|---|
| Iron nail | | | | | |
| Paraffin | | | | | |
| Sugar Cube | | | | | |
| Oil | | | | | |
| Copper Penny | | | | | |

## Thinking About

Use your chart to identify each substance.

1. Which substance is attracted to a magnet?

_______________

2. Which substance is a shiny orange-brown color and sinks in water?

_______________

3. Which substance is soluble (dissolves) in water?

_______________

4. Which substance is a solid and floats on water?

_______________

5. Which substance is not a solid and floats on water?

_______________

## Dig Deeper

## Stumper's Corner

1. ______________________________________________
   ______________________________________________

2. ______________________________________________
   ______________________________________________

## What Did You Learn ?

1. What are physical properties of chemical substances?
   ______________________________________________

2. When scientists want to know what chemical substances are in an item, they seldom consider the size, shape, and amount of the item. Why is that? ______________________________

3. Give ten examples of physical properties used by scientists to describe a chemical substance.
   ______________ ______________
   ______________ ______________
   ______________ ______________
   ______________ ______________
   ______________ ______________

4. What is a pure chemical substance? ______________________
   ______________________________________________

5. What are some of the things students learn about in analytical chemistry? ______________________________
   ______________________________________________

6. What are some of the main things that are done in medical labs?
   ______________________________________________
   ______________________________________________

7. How might an environmental agency use a lab that analyzes chemical substances? ______________________________
   ______________________________________________

8. Are the physical properties of a piece of pure iron the same any where pure iron is found? ______________________________

ACTIVITY 2

*Investigation #2*

# *Strange Substances and Their Properties*

## Thinking About

**Date:**

## The Activity: Procedure and Observations

**Part I**

Observe your bag of MX. Hold the bag by the different corners. Does it have properties of a liquid? ______________________

Hit the bag of MX (not too hard). Does it feel like a solid? ____________

Pour the contents of the bag into a plastic bowl. Pick up some MX in a spoon and let it fall back into the bowl. Does the substance act like a liquid or a solid as it falls? Describe how it falls. ________________

______________________________________________

Now slowly push your finger into the MX until your finger is touching the bottom of the pan. Pull your finger out slowly. What happened?

______________________________________________

Slowly push your finger into the MX again. When it is touching the bottom of the pan, try to pull your hand out quickly. What happened?

______________________________________________

Now try to quickly jab the surface of the MX with your fingers. What happened? ______________________________________

Try pushing the back of a spoon through a container of MX. Move the spoon as fast as you can. Describe what happens. ______________

______________________________________________

Now move the spoon through the MX very slowly. Is there a difference in how hard it is to push the spoon? ____________________

**Part II**

Will the diaper hold 50 mL of warm water? ____________________

Predict how much warm water you think the diaper can hold before it begins leaking. ________________________________________

Continue to add 50 mL of warm water until the diaper can no longer hold any more water and it steadily leaks. Record the total amount of water you added before it began to leak. ____________________

What was the total amount of the water-absorbing layer (while dry) you collected from the diaper? __________________________

Put the water-absorbing material back into the large plastic bag, and add 50 mL of warm water. What do you see? ________________

______________________________________________

What was the total amount of water you added to the bag? __________

Place the first diaper in a plastic bowl and pull it apart. Compare the inner contents of this material to the material in the gallon zip bag.

______________________________________________

Estimate how much water was added for every 100 mL of dry diaper material. This doesn't need to be exact — just an estimate. __________

List some of the physical properties of the water-absorbing chemical in the baby diaper.

________________ ________________
________________ ________________
________________ ________________

## What Did You Learn ?

1. Give several physical properties of MX.

2. There are several ways to describe viscosity. Find two or more ways to describe viscosity.

3. Viscosity of oils and molasses is often affected by temperature. What affects the viscosity of MX?

4. What is one unusual property of the chemical we tested in the baby diaper?

5. What are polymers?

## Dig Deeper

## Stumper's Corner

1.

2.

ACTIVITY 3

*Investigation #3*

# *Chemistry Fun with Bubbles*

Thinking About

Date:

## The Activity: Procedure and Observations

1. Describe the size of the bubbles.

______________________________

______________________________

______________________________

2. How far do they travel before they burst?

______________________________

______________________________

______________________________

3. Observe the colors of the bubbles. Notice what color they turn just before they burst.

______________________________

______________________________

______________________________

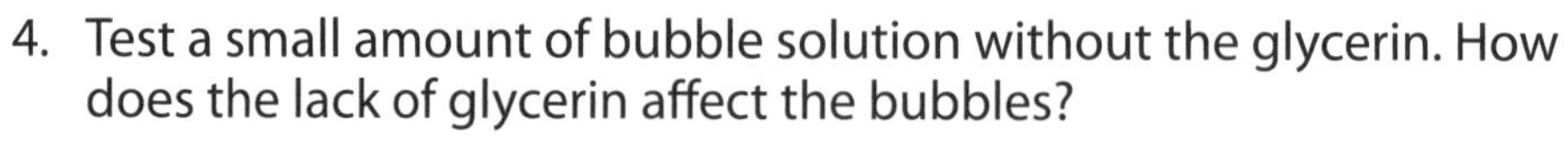

4. Test a small amount of bubble solution without the glycerin. How does the lack of glycerin affect the bubbles?

______________________________

______________________________

______________________________

## Dig Deeper

## What Did You Learn ?

1. Generally, is the attraction between molecules greater in solids or in liquids? ____________________

2. Generally, is the attraction between molecules greater in liquids or in gases? ____________________

3. The attraction between molecules that are found at the surface of a liquid is called what? ____________________

4. What is the property of matter that causes like molecules to attract each other? ____________________

5. What is the property of matter that allowed the bubbles to stretch without breaking (up to their limits)? ____________________

6. What is the difference between hard water and soft water? ____________________

7. Are the spherical shapes of bubbles caused more by surface tension, adhesion, or friction? ____________________

## Stumper's Corner

1. ____________________

2. ____________________

ACTIVITY 4

*Investigation #4*

# Colors Are Colors

## Thinking About

Date:

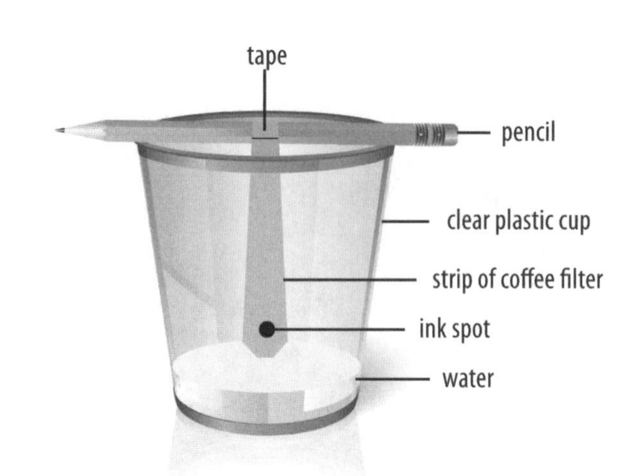


## The Activity: Procedure and Observations

Hang all strips and observe for a few minutes. Remove strips from the water and blot on paper towel. Describe what you observed the colors doing?

| Colors Tested | Results (Colors seen after test) |
|---|---|
| Black | |
| Red & Blue | |
| Blue & Yellow | |
| Red, Blue & Yellow | |

Make sure you record the original color and tell what happens to each color. Note how many separate colors you see on each strip.

## What Did You Learn ?

1. Is paper chromatography used to separate mixtures or compounds?
______________________________

2. Were some of the dyes carried up the paper moving faster than others? ____________

3. Do the chemicals in a mixture keep their own properties?
______________________________

4. If two samples of ink produce the same chromatography pattern and colors, what would this indicate?
______________________________
______________________________
______________________________

5. Suppose a chromatograph was made from a colored marker, and the pattern showed a blue spot above a pink spot. Does this give you a good idea that there are at least two chemicals in the colored marker? ____________

## Dig Deeper

## Stumper's Corner

1. ______________________________
______________________________

2. ______________________________
______________________________

**Activity 5**

*Investigation #5*

# How in the World Can You Separate a Mixture of Sand and Salt?

Date:

## The Activity: Procedure and Observations

1. Describe your observations of a mixture of sand and salt.
2. What are at least two properties that sand and salt have that are different?
3. Can you see anything in the liquid after the sand has been filtered out?
4. Lay the filter paper and sand flat on the table to dry. When it is dry, examine it with a hand lens. Does the dry sand look like it did when you observed it before? ______
5. Take a small amount of the filtered liquid and pour it into a shallow glass container. What do you see in the dish after the liquid has evaporated?

## Thinking About

6. Examine the recrystallized salt with a hand lens. What do you observe?

## Dig Deeper

## What Did You Learn ?

1. Which of the following are examples of mixtures: salt and sand stirred together, crude oil, salt water, distilled water.

____________________

2. What is one way a mixture is different from a pure substance?

____________________

3. When two or more pure substances are mixed together, do they keep their individual properties? ____________

4. When two or more pure substances are combined chemically, do they keep their individual properties? ____________

5. Distillation is a way of separating mixtures of liquids. This process depends on differences in what physical property?

____________________

6. What kind of substance can be separated from a liquid by a funnel and filter paper — one that is dissolved in the liquid or one that is not dissolved in the liquid? ____________

7. How can you separate a mixture of salt and water?

____________________

## Stumper's Corner

1. ____________________

2. ____________________

## Quiz ?

1. Do you know the first time the word water is mentioned in the Bible?

____________________

2. What did God do with the waters on the second day of creation?

____________________

3. What did God do with the waters on the third day?

____________________

4. What did God do to the water on the fifth day?

____________________

5. To whom did God ask this question: "Who shut up the sea behind doors when it burst forth from the womb . . . when I fixed limits for it and set its doors and bars in place . . . ?" ____________

6. What great writer in the Bible makes the observation that all the rivers run into the ocean, but the ocean never gets full?

____________________

7. To whom did Jesus tell " If you knew the gift of God and who it is that asks you for a drink, you would have asked him and he would have given you living water." ____________

8. Who said, "Whoever drinks the water I give him will never thirst."

____________________

9. Peter wrote (2 Peter 3:6) that "the world that was" perished. How did this happen?

____________________

ACTIVITY 6

# *Investigation #6*
# *Water Is the Standard*

**Date:**

## The Activity: Procedure and Observations

1. Show your partner the object you described. How accurate was your partner's drawing? ____________

2. Look at your drawing. How accurately did you draw your partner's object? ____________

3. What did the water level rise to when you lowered a rock into the graduate? ____________

4. The difference in this level and the 30 mL of water you started with is the volume of the rock. What is the volume of the rock in mL? (You may change mL to $cm^3$. See "The Science Stuff.") ____________

5. Draw a line that is 3⅝ inches long.
   What is the volume of a cubic block that is 3⅝ inches on each side?

   ____________

6. Draw a line that is 3.7 cm long.
   Measure the height, length, and width of the rectangular block you are given. ____________

   height = ______cm    length = ______cm    width = _______cm

## Thinking About

7. What is the volume of the wooden block? ____________

8. Did you make a cubic centimeter paper box? ____________

9. If the box was filled with iron, what would it weigh? ____________

10. If it were filled with wood (approximately, since all pieces of wood are not the same), what would it weigh? ____________

11. If it were filled with ethyl alcohol, what would it weigh? ____________

12. Find three more substances on the density chart. Tell how much each would weigh if they exactly filled your cubic centimeter box.

    ____________
    ____________
    ____________

The correct reading is 13mL. You read the bottom of the curved surface of the water and other liquids.

## Dig Deeper

## What Did You Learn

1. How would you find the volume of a small, irregularly shaped piece of metal? ______________________

______________________

2. What happens to a person's mass as the distance from the earth increases? ____________ What happens to the person's weight?

______________________

3. Give an example of a metric unit that is used to measure volume.

______________________

4. Give an example of a metric unit that is used to measure how long something is. ______________________

5. Give an example of a metric unit that is used to measure an object's mass. ______________________

6. What is the density of pure water? Write this as words and as symbols.

______________________

7. If a substance has a density of 2.5 g/mL, will a block of this substance float or sink in water? ______________________

8. How can you calculate the density of a substance? ____________

______________________

9. How much does 25 mL of water weigh? ______________________

10. How can you find the volume of a large rectangular wooden block?

______________________

______________________

11. The displacement of water method is used to find the volume of a rock. Its volume is found to be 19 mL of water. What is the volume of the rock in $cm^3$? ______________________

______________________

## Stumper's Corner

1. ______________________

______________________

2. ______________________

______________________

ACTIVITY 7

*Investigation #7*

# *Bending Streams of Water*

## The Activity: Procedure and Observations

Date:

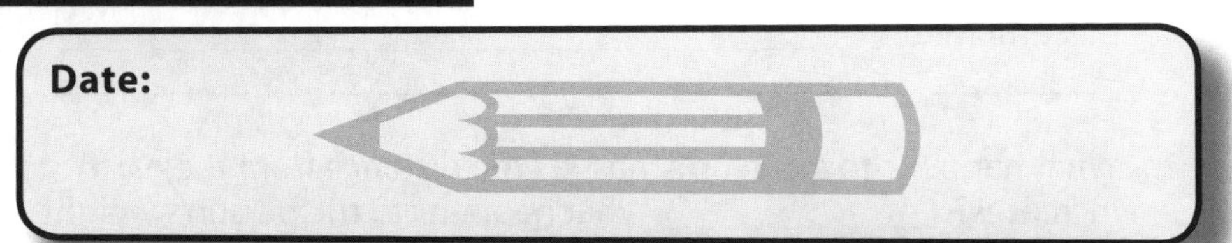

1. Do you see some of the paper moving as you moved a charged balloon over it?

2. Touch the balloon firmly with your hand. Move the balloon over the paper again. Does anything happen this time? ___________________

3. Describe how a stream of water is affected by a charged balloon.

   ___________________________________________________

4. Draw a picture of what you observed with the balloon and the stream of water.

**Drawing Board:**

## Thinking About

5. After you have pushed toothpicks into a Styrofoam ball from the four corners of a tetrahedron, observe where they are located. Are any two toothpicks on opposite sides of the ball? _________________

6. Make a model of a molecule of water using one large Styrofoam ball, two small balls, and four toothpicks. Draw a picture of your water model.

**Drawing Board:**

7. Rearrange the toothpicks and two small balls according to the directions so that the two small balls are on opposite sides of the large ball. Do you see the difference between this arrangement and the tetrahedron arrangement? _________________________________

8. Draw a picture of this arrangement and label it "not a water molecule."

**Drawing Board:**

9. Describe at least four physical properties of water that you can observe with your senses.

______________________________________________

______________________________________________

______________________________________________

______________________________________________

## Dig Deeper

## What Did You Learn ?

1. All matter is made up of what kinds of charges? ______________
2. Name some things that can be dissolved in water. ______________
3. A water molecule is made up of which two kinds of atoms? ______________
4. What geometric shape explains one reason why water molecules are polar? ______________
5. What are the molecules called that have strong connecting bonds, a positive charge on one end, and a negative charge on the other end? ______________
6. List at least four physical properties of water. ______________
7. Write the formula for water and explain each symbol and number. ______________
8. What is the difference in an atom and a molecule? ______________

## Stumper's Corner

1. ______________________________________________
2. ______________________________________________

ACTIVITY 8

*Investigation #8*

# *Drops of Water*

Date:

## The Activity: Procedure and Observations

1. How do the five drops of water compare in size?

2. Put the sharpened point of your pencil into a drop of water and describe how the water responds when the pencil is lifted out.

3. Touch the eraser of your pencil to a drop of water and describe how the water responds when the eraser is lifted out.

4. Are the water molecules more attracted to the lead end or the rubber end? ______

## Thinking About

5. With the point of the pencil, push one drop around the paper. Next, push two drops together, and then three and four. Pay attention to the details and write your observations.

6. What happens to the drops of water when soap is added?

7. Predict how many total drops of water you can place on a penny before it overflows.

8. Now test your prediction by putting one drop of water at a time on a clean penny with a medicine dropper. Record the number of drops that will sit on a penny before some of the water overflows.

9. Repeat the experiment, but this time use one of the following solutions — salt water, soapy water, or rubbing alcohol and water. Count the number of drops of the second solution you can place on a penny before it overflows.

## Dig Deeper

## Stumper's Corner

1. ______________________________________________

______________________________________________

2. ______________________________________________

______________________________________________

## What Did You Learn ?

1. One end of a water molecule is positive and one end is negative. What happens when the positive end of one molecule comes near the negative end of another molecule? ______________ Make drawings to illustrate this.

**Drawing Board:**

2. What is a polar molecule? ______________________________

______________________________________________

3. At the surface of the water, the molecules attract each other in every direction except up. What name is given to the attraction of water molecules at the surface? ______________________

4. Why do water molecules have an attraction for each other? ________

______________________________________________

5. Adding soap or detergent to a drop of water causes the drop to flatten out. What does soap break down when it is added? __________

______________________________________________

6. To correctly read the amount of water in a graduated cylinder, what part of the curved shape of the surface of the water must one read? ______________________ Make a diagram if you wish.

7. Is there an attraction between the water and the glass (or plastic) in a graduated cylinder? _______ Is this attraction known as cohesion or adhesion? __________

8. If you divide the mass of a substance by its volume, what would you calculate? ______________________________

9. If several drops of water are placed on a clean penny, what kind of shape will the water have? ________________ Why is this? ________

______________________________________________

**ACTIVITY 9**

## *Investigation #9*
# *Oil and Water Don't Mix*

Date:

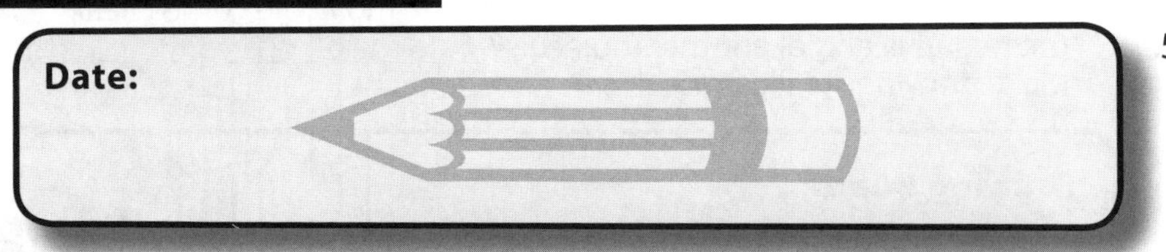

## The Activity: Procedure and Observations

1. Pour equal amounts of oil and water into the cup (#1). Write your observation.

   ______________________________

   ______________________________

   ______________________________

2. Stir the mixture with a stick and make further observations.

   ______________________________

   ______________________________

   ______________________________

3. Pour equal amount of oil and water into the container (#2), close the lid, and shake several times. Write your observations.

   ______________________________

   ______________________________

   ______________________________

4. Add a few of drops of soap to the cup (#1) and stir well. Did the oil and water mix in the cup when stirred? ______________

## Thinking About

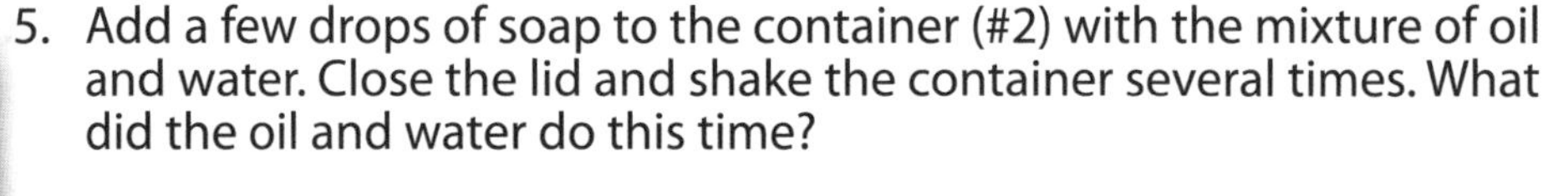

5. Add a few drops of soap to the container (#2) with the mixture of oil and water. Close the lid and shake the container several times. What did the oil and water do this time?

   ______________________________

   ______________________________

   ______________________________

6. What difference did the addition of a few drops of soap make to the mix? ______________

7. At the beginning of the lesson, you rubbed a fatty substance on your hand and tried to wash it off with water. How easily did the fat come off your hands? ______________

8. Now rub a fatty substance on your hands as you did before. Wash your hands with soap and water. Did the fatty substance come off easily this time? ______________

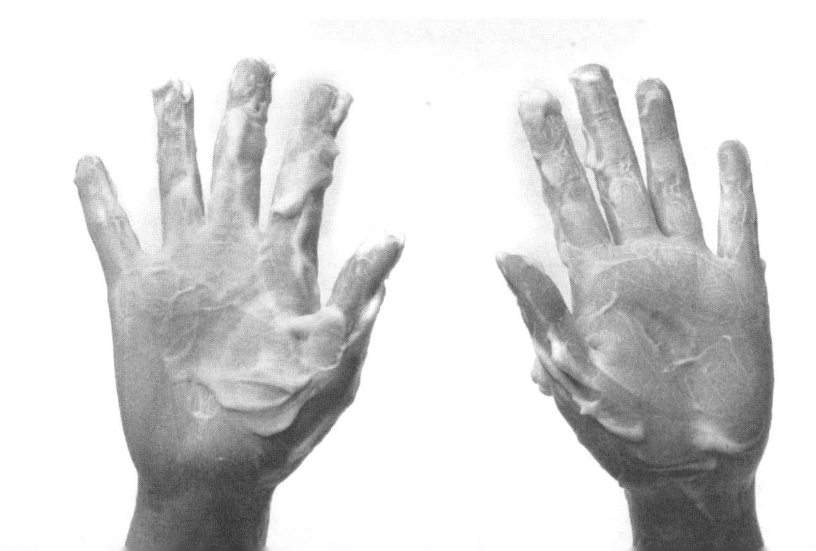

## Dig Deeper

## What Did You Learn

1. Some of the most important properties of water occur because water molecules are slightly positive at one end and slightly negative at the other end. What are these kinds of molecules called?
   ______________________________
2. Are oil molecules polar or non-polar? ______________
3. What can you generally predict about dissolving non-polar substances in polar substances? ______________
   ______________________________
4. How is soap able to dissolve both polar and non-polar substances?
   ______________________________
   ______________________________
   ______________________________
5. Why is it hard to get oily substances, such as lipstick, out of clothing using just water?
   ______________________________
   ______________________________
6. Why is it hard to wash oil off your hands only using water?
   ______________________________
   ______________________________

## Stumper's Corner

1. ______________________________
   ______________________________
2. ______________________________
   ______________________________

## Narrative: Is There Life on Other Planets

**Discuss**: What are some of the ways the earth was created that allow living things to make their homes on it?

ACTIVITY 10

*Investigation #10*

# *Acids and Bases*

Date:

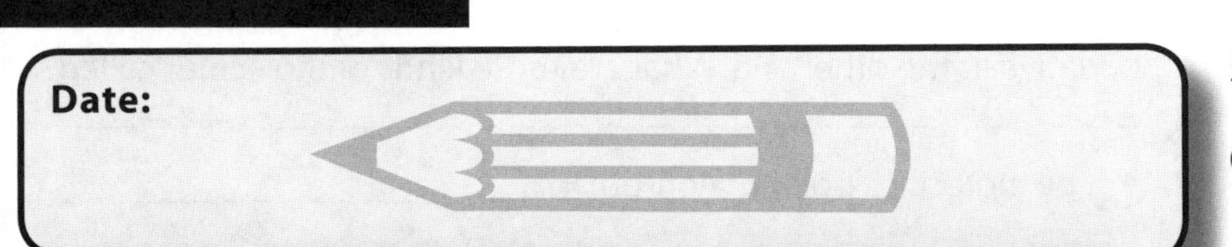

## The Activity: Procedure and Observations

1. Taste a few of the pieces of some citrus fruits. The fruits you tasted all contained acids. What is a property related to taste that they all have in common?

   ______________________________

   ______________________________

2. Give examples of some other foods we eat that have a sour taste.

   ______________________________

   ______________________________

3. Your teacher will give you some household bases to feel. (Don't taste this.) What was a common property of the bases related to how they feel?

   ______________________________

   ______________________________

4. Give some examples of other things that feel slippery.

   ______________________________

   ______________________________

## Thinking About

5. Taste the water. Does it have a sour taste? ______________

6. Feel the water. Does it have a slippery feel? ______________

7. From what you have learned, do you think water is an acid, a base, or neutral? ______________

8. Follow the direction for testing items with litmus paper. What color does litmus paper turn in fruit juice? ______________

9. What color does litmus paper turn in the soaps and cleaners?

   ______________________________

10. What did you observe about the red litmus paper in the water?

    ______________________________

    ______________________________

11. What did you observe about the blue litmus paper in the water?

    ______________________________

    ______________________________

12. Does this test make you more sure or less sure about your earlier decision of whether water is an acid, a base, or neutral?

    ______________________________

    ______________________________

## Dig Deeper

## What Did You Learn

1. Do citrus fruits contain acids or bases? ____________________
2. Do many common cleaners contain acids or bases? ____________
3. Are acids and bases found in water solutions or in oily solutions?

   ____________________________________________
4. What happens to acids and bases when they are in water solutions?

   ____________________________________________
5. Acids usually have what kind of taste? ____________________
6. Bases usually have what kind of taste and what kind of feel?

   ____________________________________________
7. Chemicals that change color in acids and bases are called what?

   ____________________________________________
8. What color does litmus paper turn in acids? ________________
   In bases? ________________________
9. What pH numbers indicate an acid? ________ A base? ________
   A neutral substance? ________________________
10. Are positively charged hydrogen atoms found in acids or in bases?

    ______________

## Stumper's Corner

1. ____________________________________________

   ____________________________________________

2. ____________________________________________

   ____________________________________________

ACTIVITY 11

*Investigation #11*

# *Basically – Is It Acid or Base?*

## Thinking About 

Date:

## The Activity: Procedure and Observations 

Name each chemical you will be testing. Predict whether each chemical is an acid or a base. Observe the color the red cabbage juice turns when each chemical is added to it. Put this information in the data table.

| Chemical | Predict: A or B | Color change | Estimated pH |
|---|---|---|---|
| | | | |
| | | | |
| | | | |
| | | | |
| | | | |
| | | | |
| | | | |
| | | | |
| | | | |
| | | | |

1. Use the pH chart from The Science Stuff to estimate the pH number of each chemical and add this information to the data table.

2. Use the information in the data table to classify the chemicals you used as strong acid, weak acid, neutral, weak base, or strong base.

3. Find the cup with red cabbage juice and ammonia. Add one drop of vinegar at a time to the ammonia/cabbage juice. Count the drops until there is a color change. What color change occurred?

    ______________________________

    ______________________________

4. Find the cup with red cabbage juice and distilled water. Blow into the liquid with a straw for a few minutes. Do you observe any change in the color of the cabbage juice? ______________________________

pH Chart

| Reds | Purples | Violet | Blue | Blue-green | Greenish-yellow |
|---|---|---|---|---|---|
| 2 | 4 | 6 | 8 | 10 | 12 |

## Dig Deeper

## What Did You Learn ?

1. Do all soils have the same pH level? ___________
2. What is an indicator? ___________
3. Give some examples of indicators. ___________
4. Can some indicators tell the difference between a strong acid and a weak acid? ___________
5. If you accidentally spilled a strong acid or base on your skin, what is the first thing you should do? ___________
6. What new chemical forms when there is a chemical reaction between carbon dioxide and water? ___________
7. What new kinds of chemicals form when an acid and a base react chemically? ___________
8. Is ordinary rainwater neutral, slightly acidic, or slightly basic? ___________
9. Acid rain may be produced when water in the atmosphere reacts with what kinds of air pollutants? ___________
10. What would a pH number of seven tell you about a chemical? ___________ What would a pH of two indicate? ___________

## Stumper's Corner

1. ___________
2. ___________

ACTIVITY 12

*Investigation #12*

# *Salt — An Ordinary Substance with Extraordinary Powers*

**Date:**

## The Activity: Procedure and Observations

1. Examine a sample of table salt crystals with a magnifying lens. Describe the shape of the crystals (don't worry about the size). Describe other properties of the crystals, such as color and texture.

_______________
_______________
_______________

2. Use the back of the spoon to press on some of the crystals. Are they easily crushed? _______________

3. After a tablespoon of table salt has been completely dissolved in a glass of water, observe carefully to see if there is any evidence that the salt is in the water.

   Allow the water to evaporate from a teaspoonful of the salt water on a shallow container. Use the magnifying lens to examine the crystals that form. How do the properties of these crystals compare to the salt crystals you observed earlier?

_______________
_______________
_______________

## Thinking About

4. Look at a copy of the Periodic Table of the Elements. Is table salt listed on the Table? _______________

5. Where do you find metals on the Periodic Table?

_______________

6. Where do you find nonmetals on the Periodic Table?

_______________

7. Are there more kinds of metals or nonmetals on the Periodic Table?

_______________

8. Use the Periodic Chart to tell if the following elements are a metal or a nonmetal: rubidium __________, cadmium __________, sulfur __________, niobium __________, phosphorus __________, barium __________, and neon __________.

9. See if you can name these compounds:

   KCl _______________

   LiF _______________

## Dig Deeper

## What Did You Learn

1. Does the Periodic Table contain elements or compounds?

   ______________________________

2. Where are metals found on the chart? ______________________________
3. Where are nonmetals found on the chart? ______________________________
4. What kinds of elements are generally found in a salt?

   ______________________________

5. What is a chemical symbol? ______________________________
6. What is a chemical formula? ______________________________
7. Write the chemical formula for sodium chloride and tell what information the formula gives you.

   ______________________________

8. List at least three physical properties of sodium chloride.

   ______________________________

   ______________________________

9. When an acid and a base combine in the right amounts, what do they change into? ______________________________
10. Why don't you find water listed on the Periodic Table?

    ______________________________

11. Do sodium and chlorine keep their same properties when they chemically combine to make salt? ______________________________

## Stumper's Corner

1. ______________________________

   ______________________________

2. ______________________________

   ______________________________

## Pause and Think: The Salt of the Earth

What were your thoughts about this?

______________________________

______________________________

ACTIVITY

13

## *Investigation #13*
# *More About the Amazing Periodic Table*

## Thinking About

**Date:**

## The Activity: Procedure and Observations

1. Predict four salts that could result from combinations of elements from columns IA and VIIA. Name them and write their formulas using sodium chloride as your example.
   1.
   2.
   3.
   4.

2. Predict four salts that could result from combinations of elements from columns IIA and VIIA. Name them and write their formulas using calcium fluoride as your example.
   1.
   2.
   3.
   4.

3. Give an example of one element from each of the following columns:

   Column VIIIA ____________________

   Column IA ____________________

   Column VIIA ____________________

4. See if you can predict which element in each pair would be heavier (assume equal sizes):

   iron or gold __________

   lead or actinium __________

   aluminum or tin __________

   sodium or potassium __________

   hydrogen or helium __________

5. Use the Periodic Table and predict what might have happened if the Hindenburg had been filled with helium gas instead.

   ____________________________________________

   ____________________________________________

   ____________________________________________

6. Find the dividing line between metals and nonmetals. Recall the general properties of metals and nonmetals.

   Predict four elements that would conduct electricity:

   __________, __________, __________, __________

   Predict four elements that would not conduct electricity:

   __________, __________, __________, __________

## Stumper's Corner

1. ____________________________________________

   ____________________________________________

2. ____________________________________________

   ____________________________________________

## Dig Deeper

## Pause and Think: Metals in the Bible

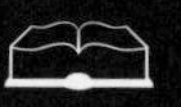

Use a Bible Concordance and try to find how many metals are mentioned in the Bible. Name the metal and give at least one reference to that metal.

______________________________

______________________________

______________________________

## What Did You Learn ?

1. The Periodic Table consists of a series of blocks containing symbols and numbers organized into columns and rows. What do these blocks contain information about — elements or compounds?

   ______________________________

2. Where do you find similar groups of elements — in vertical columns or in horizontal rows? ______________

3. Except for hydrogen, do you find metals or nonmetals on the left side of the Table? ______________

4. Where are most nonmetals found?

   ______________________________

   ______________________________

5. Predict: Which element in each pair would be more dangerous or more reactive — potassium or calcium; sulfur or chlorine; krypton or selenium? ______________

6. Predict: Would astatine have properties more like radon or like iodine? ______________

7. Which scientist organized the known elements into a Periodic Table and left blanks where he predicted undiscovered elements would go? ______________

8. Predict: Would a block of gold weigh more or less than an equal-size block of uranium? ______________

9. Predict: Which element would be the best conductor of electricity — palladium or phosphorus? ______________

10. Predict: Which element would be a shiny solid — rhenium or krypton? ______________

11. What does an acid and base change into when they react chemically?

    ______________________________

    ______________________________

ACTIVITY

14

## *Investigation #14*
# *Electricity and Salt Water*

**Thinking About**

**Date:**

## The Activity: Procedure and Observations

Set up the apparatus as described in your book.

Record your observations. Note which electrode has the most bubbles coming off of it.

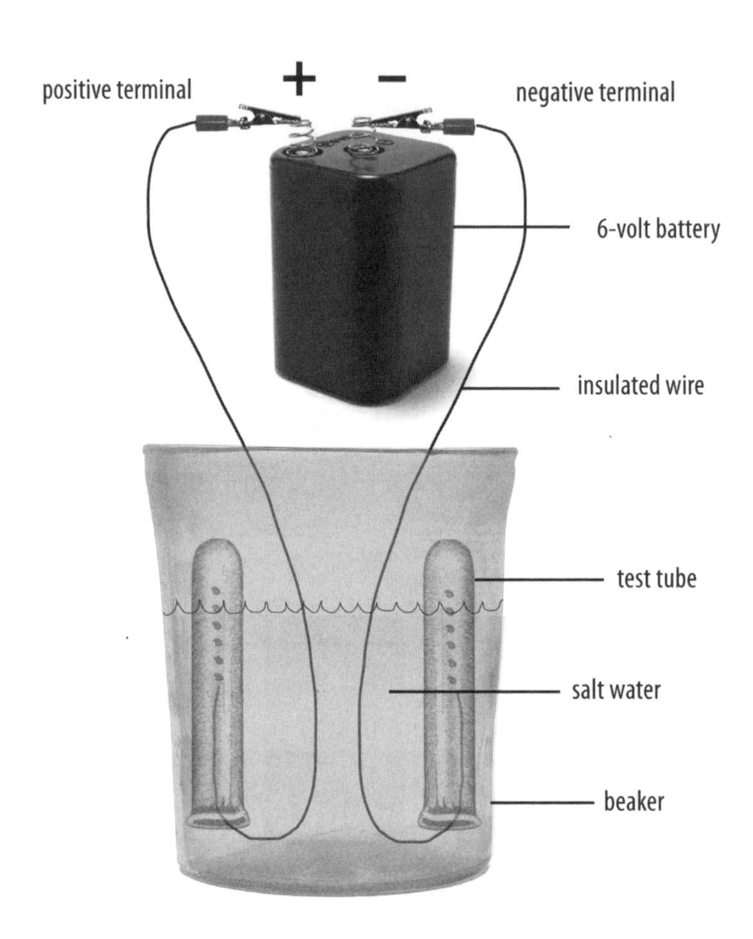

## What Did You Learn ?

1. What is electrolysis?

_______________________________________________

_______________________________________________

2. When water is separated by electrolysis, what gas forms at the positive electrode? ______________________

3. When water is separated by electrolysis, what gas forms at the negative electrode? ______________________

4. Are there more hydrogen or oxygen bubbles formed? __________

5. What was the purpose of adding the salt?

_______________________________________________

_______________________________________________

6. Will pure water conduct an electric current? ______________

7. Suppose someone in the 1500s had invented some method to separate water into oxygen and hydrogen gases. This would have been strong evidence against what popular theory?

_______________________________________________

## Dig Deeper

## Stumper's Corner

1. _______________________________________________

_______________________________________________

2. _______________________________________________

_______________________________________________

ACTIVITY 15

*Investigation #15*
# *Changes — Are They Chemical or Physical?*

**Thinking About** 

**Date:**

## The Activity: Procedure and Observations

**Activity 1**

Record all the changes you observe as sugar is heated. Note the smoke and any color changes. When the black, puffy material is cool, feel it to determine its texture and strength.

Record all your observations. Be as detailed as you can and use your senses of sight, smell, and touch.

Try to combine a little water and carbon in another container and see if you can make water and carbon molecules go back together to remake sugar. What happens?

**Activity 2**

Observe a mixture of sugar and sand with a hand lens and record your observations.

Can you identify the grains of sugar? How are they different from grains of sand?

Separate a few of each kind of grains into two groups. Can you crush either of them?

Add a few drops of water with the spoon to both groups. Will either of them dissolve in water?

**Activity 3**

Compare the properties of the sugar and water mixture to the properties of pure water.

## Dig Deeper

## Stumper's Corner

1. ______________________________________________

______________________________________________

2. ______________________________________________

______________________________________________

## What Did You Learn ?

1. Give several examples of physical changes.

______________________________________________

______________________________________________

2. Give several examples of chemical changes

______________________________________________

______________________________________________

3. Read the definitions of both physical and chemical changes. Tell the differences in your own words.

______________________________________________

______________________________________________

4. Name the kinds of atoms that make up a sugar molecule.

______________________________________________

______________________________________________

5. What is the black substance that is left after sugar decomposes?

______________________________________________

6. What was the name of the white smoke that formed from hydrogen and oxygen atoms released from the sugar?

______________________________________________

7. Was the decomposition of sugar a chemical or a physical change?

______________________________________________

8. Is dissolving sugar in water a chemical or a physical change?

______________________________________________

9. Give two ways in which sugar and sand have different physical properties.

______________________________________________

______________________________________________

ACTIVITY 16

*Investigation #16*

# *Clues of a Chemical Reaction*

## Thinking About

**Date:**

## The Activity: Procedure and Observations

**Part 1**

1. As much as you can, observe evidence of bubble formation and color changes. Note any sounds your hear as well. After a few minutes, remove the lid to the cup with the Phenol Red. What color is the solution now? ____________________

2. Can you tell that changes occurred in both test tubes? Record your observations.

____________________

____________________

____________________

3. What clue would make you think a chemical reaction had taken place in the cup with the Phenol Red?

____________________

____________________

____________________

**Part 2**

As much as you can, observe evidence of bubble formation and color changes. Note any sounds you hear as well. After a few minutes, remove the lid to the cup with the Phenol Red. What color is the solution now?

____________________

Can you tell that changes occurred in both test tubes? Record your observations.

____________________

____________________

____________________

What clue would make you think a chemical reaction had taken place in the cup with the Phenol Red? ____________________

**Part 3**

Your teacher will give you two liquids: one produced by soaking a steel-wool pad in vinegar overnight and one made by steeping a tea bag in hot water. Record your observations of the two liquids.

____________________

____________________

____________________

Mix a little of each solution in a clear plastic cup. What clue would make you think a chemical reaction had taken place? (Caution: Be careful when handling this. It is like permanent ink that doesn't wash out of clothing!) ____________________

**Part 4**

What happens when liquid dishwaher detergent and a solution of Epson salts are combined?

______________________________________________

______________________________________________

## Dig Deeper

## What Did You Learn ?

1. Which of these changes are chemical changes: boiling water, freezing water, adding vinegar to baking soda, dissolving salt in water, combining an acid and a base.

______________________________________________

______________________________________________

2. A forensic scientist wants to know if a certain chemical is present. When a few drops are added to a solution, an insoluble substance forms that is a bright yellow. Is it likely that a chemical change occurred? ______________________

3. What is the difference in a physical change and a chemical change?

______________________________________________

4. A piece of zinc is added to a test tube containing hydrochloric acid. Bubbles start to rise in the tube and the tube begins to feel warm. Is it likely that a chemical change occurred? ______________

5. Suppose you have an unknown gas and you allow a small amount of the gas to bubble through a solution of limewater. What would happen if the gas were carbon dioxide?

______________________________________________

______________________________________________

6. What are four clues that a chemical reaction has taken place?

______________________________________________

______________________________________________

______________________________________________

______________________________________________

## Stumper's Corner

1. ______________________________________________

______________________________________________

2. ______________________________________________

______________________________________________

ACTIVITY 17

## *Investigation #17*
# *A Heavy Gas*

**Date:**

## The Activity: Procedure and Observations

1. Set up the apparatus as you are instructed. Begin observing what happens when the baking soda and vinegar are combined. Include measurements of the balloon and the time the reaction lasted.

___________________________________________
___________________________________________
___________________________________________

2. What happens when a drinking glass is lowered over a burning candle?

___________________________________________
___________________________________________
___________________________________________

## Thinking About

3. What happens to a burning candle when bubbles are released around it from vinegar and baking soda?

___________________________________________
___________________________________________
___________________________________________

4. When you finish, list as many properties of the gas that formed as you can.

___________________________________________
___________________________________________
___________________________________________

## Dig Deeper

## Stumper's Corner

1. ____________________________________________
____________________________________________

2. ____________________________________________
____________________________________________

## What Did You Learn ?

1. What two things did you observe that would make you think a gas was being produced in the activity you did?
____________________________________________
____________________________________________

2. What was the name of the gas that was produced when vinegar and baking soda reacted?
____________________________________________
____________________________________________

3. In order for any substance to burn, what gas must be present?
______________________________

4. List three properties of carbon dioxide.
____________________________________________
____________________________________________
____________________________________________

5. What is "dry ice"?
____________________________________________
____________________________________________

6. Does "dry ice" melt when it gets warmer like ordinary ice does?
____________________________________________

7. Explain why a burning candle will go out when carbon dioxide is produced around it.
____________________________________________

8. Explain why a burning candle will go out when a glass is placed over it.
____________________________________________

9. A green plant takes in and uses what gas as it makes food?
____________________________________________

10. Is carbon dioxide a polar compound? ____________________

ACTIVITY 18

*Investigation #18*

# *Large or Small? Hot or Cold?*

## Thinking About

**Date:**

## The Activity: Procedure and Observations

**Part 1**

Place labels under the three cups to identify effervescent tablet as fine powder, small pieces, and a whole piece. Add 40 mL of room-temperature water to each cup. Begin timing as soon as the water is added. Record the time it takes for the bubbles to stop forming in the following chart.

| | Reaction time |
|---|---|
| Fine powder | |
| Small pieces | |
| Whole piece | |
| Which half reacted at the fastest rate? | |
| Which half reacted at the slowest rate? | |

**Part 2**

Plan your own experiment to determine the effect of temperature on reaction time, using three different water temperatures (ice water, room temperature water, and almost boiling hot water). Write your procedure.

Let someone else look at your procedure to see if you have thought of everything you needed to do. Who was your reader?

Record the time for the reactions in a chart.

| | Reaction time |
|---|---|
| Ice water | |
| Room temp. water | |
| Hot water | |
| Which half reacted at the fastest rate? | |
| Which half reacted at the slowest rate? | |

There are things you kept the same (called the controls) in the three cups. What were the controls? ______________________

There is one thing that was different (called the variable) in the three cups. What was the variable? ______________________

## Dig Deeper

## Stumper's Corner

1. ______________________________

______________________________

2. ______________________________

______________________________

## What Did You Learn ?

1. Do all chemical reactions occur at the same rate? ____________
2. Explain how the reactions you observed were made to occur at a faster rate.

______________________________

______________________________

3. In order for two molecules to react chemically with each other, do they need to bump into each other? ____________
4. Why will small pieces of effervescent tablet react faster than one big piece?

______________________________

______________________________

______________________________

5. A rapid chemical reaction occurs when zinc metal is placed in a container of sulfuric acid. Will the reaction occur faster if the zinc is divided into several small pieces? ____________
6. Which would have more surface area — an apple that is cut into two halves or an apple that is cut into four quarters? ____________
7. Explain why adding heat causes most chemical changes to react faster. ____________

______________________________

8. Give an example of an industry where very high temperatures are needed to cause a chemical reaction to proceed better.

______________________________

______________________________

9. What were the controls in Part II? ____________
10. What was the variable in Part II? ____________
11. What is the procedure in an experiment? ____________

______________________________

12. What is a peer review? ____________
13. Can you do one experiment in which you test the effects of particle size and the effects of temperature at the same time? ____________

ACTIVITY 19

*Investigation #19*

# Understanding Phase Changes

## Thinking About

Date:

## The Activity: Procedure and Observations

Follow the directions for showing how close the water molecule models are in each phase — solid, liquid, and gas. Make copies of the models in the drawing.

Label the changes that occur as you move up the drawing, using the terms melting and evaporation. Now start at the top of the drawing and move down, using the terms condensation and freezing.

As you move up the page (from solid to liquid to gas), molecules absorb energy. As you move down the page (from gas to liquid to solid), molecules release energy. Write this information on your drawing.

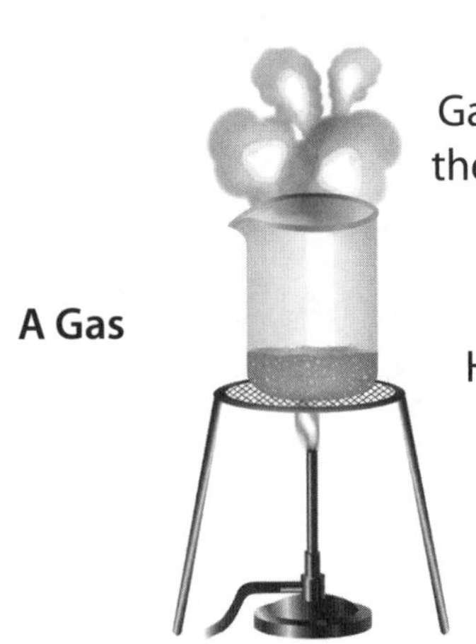

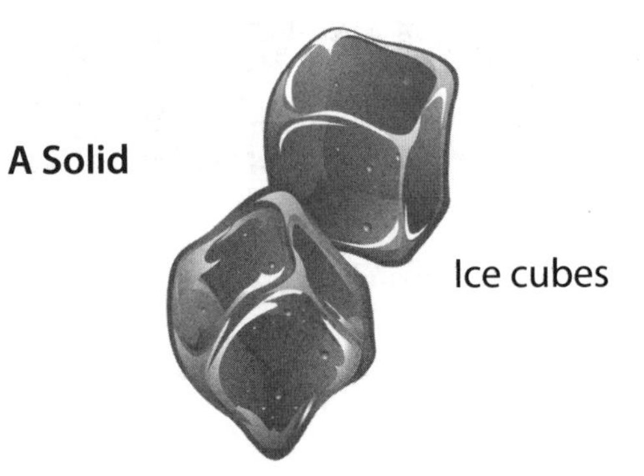

## Dig Deeper

## What Did You Learn ?

1. How close are molecules of a gas to each other compared to how close they are in a liquid? ______

2. How close do molecules of a liquid get to each other compared to how close they are in a solid? ______

3. Are phase changes physical changes or chemical changes? ______

4. What happens to the movement of molecules when heat energy is added to a substance? ______

5. What happens to the movement of molecules when heat energy is given off by a substance? ______

6. There is an important exception to the general rule that solids shrink when they are frozen. What substance gets bigger when it freezes? ______

7. Why are cement sidewalks made with cracks between the blocks? ______

## Stumper's Corner

1. ______

2. ______

**Activity 20**

*Investigation #20*

# *The Race to Evaporate*

**Date:**

## The Activity: Procedure and Observations

Label each of the four circles. Observe and note the time as soon as the substance is no longer visible. Record the time it takes for each substance to evaporate.

| | Evaporation time |
|---|---|
| Ice | |
| Acetone | |
| Rubbing alcohol | |
| Water | |

Which one evaporated the fastest? ______________________

Rank them in order from the fastest to the slowest to evaporate.

1.
2.
3.
4.

## Thinking About

Have the teacher put two drops of rubbing alcohol on the back of your hand. Now blow or fan over the alcohol until it evaporates. Does your skin feel cooler where the alcohol had been? ______________________

Put two drops of water on the back of your other hand. Now blow or fan over the water until it evaporates. Does your skin feel cooler where the water had been? ______________________

Did you notice a difference in how cold your skin felt where rubbing alcohol had evaporated and where water had evaporated? __________

## Dig Deeper

## Stumper's Corner

1. ______________________________________________
______________________________________________

2. ______________________________________________
______________________________________________

## What Did You Learn ?

1. Where did each substance go when it evaporated?
______________________________________________
______________________________________________

2. In what form was each substance after it had evaporated?
______________________________________________
______________________________________________

3. Are the substances you started with still in the room? ____________

4. What are some of the factors that affect how quickly a substance evaporates?
______________________________________________
______________________________________________
______________________________________________

5. Why does it take ice longer to evaporate than it takes an equal amount of water to evaporate? ____________

6. In your investigation, the ice was not an equal amount compared to the other substances. Why would this indicate that you had added an additional variable? ____________

7. Is evaporation a cooling or a warming process? ____________

8. Explain how sweating helps to keep your body cool if you are running on a hot day.
______________________________________________
______________________________________________
______________________________________________

9. Does evaporation occur at the surface of a liquid or below the liquid's surface? ____________

# Appendix

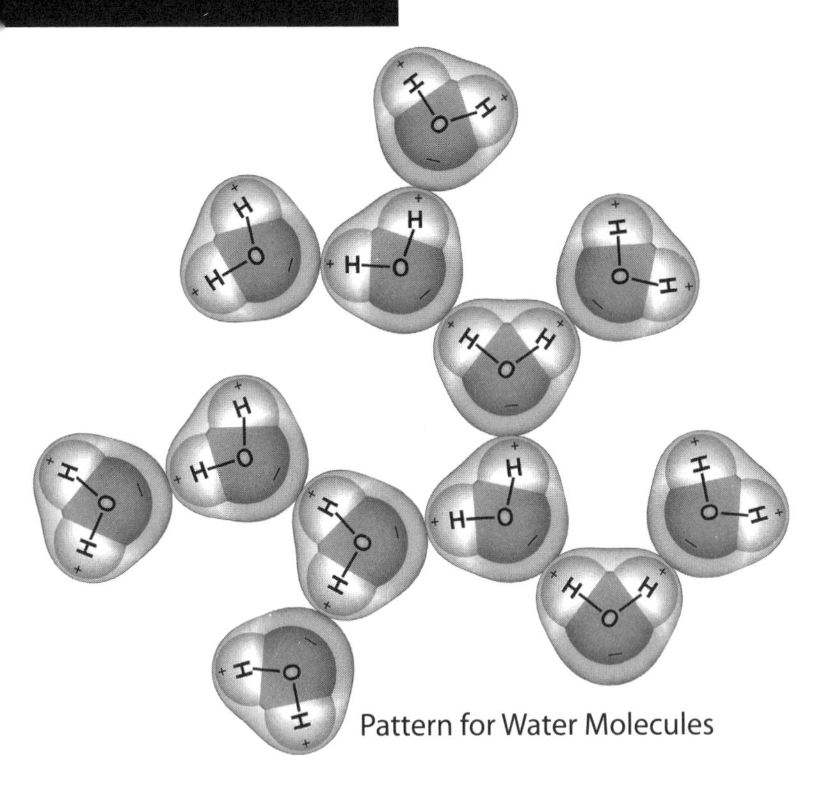

Pattern for Water Molecules

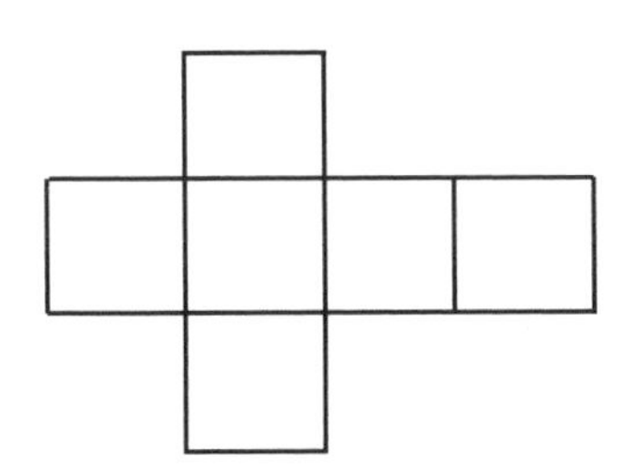

Pattern for a Cubic Centimeter($cm^3$) Box

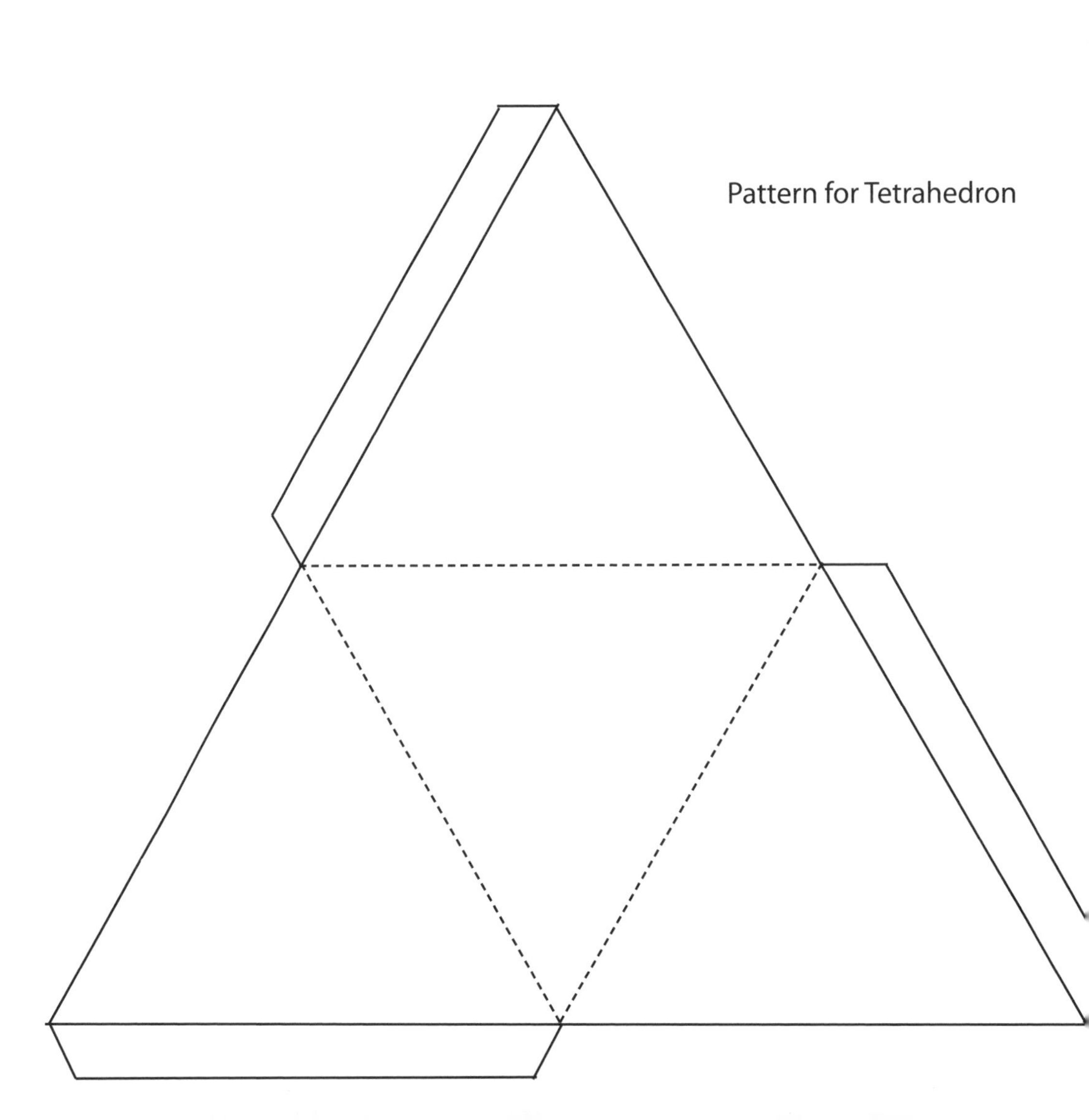

Pattern for Tetrahedron

## Additional Notes

## Additional Notes

## Additional Notes

# An exciting 3rd to 6th grade science curriculum!

Finally, a way to study science that your kids will love. Learn science basics like physics and chemistry while you have lots of fun with this activity-based series. Includes activities related to:

- friction
- speed
- inertia
- speed
- lift
- mass
- solar power
- gravity
- energy
- force
- density
- wind energy
- periodic table
- chemical reaction
- acids
- water
- heat

Important scientific terms and concepts are introduced in an engaging series of investigative lessons. Designed to fit into any education program, this informative series offers the best in science education and biblical reinforcment.

11 x 8.5 • Perfect Bound • Full Color • 88 pages
Retail: $12.99 U.S.
ISBN-13: 978-0-89051-539-6

Forces and Motion - Teacher's Guide
11 x 8.5 • Paperback • 48 pages
Retail: $4.99 U.S.
ISBN-13: 978-0-89051-541-9

Forces and Motion - Student Journal
11 x 8.5 • Paperback • 48 pages
Retail: $4.99 U.S.
ISBN-13: 978-0-89051-540-2

11 x 8.5 • Perfect Bound • Full Color • 88 pages
Retail: $12.99 U.S.
ISBN-13: 978-0-89051-570-9

Energy - Teacher's Guide
11 x 8.5 • Paperback • 48 pages
Retail: $4.99 U.S.
ISBN-13: 978-0-89051-572-3

Energy - Student Journal
11 x 8.5 • Paperback • 48 pages
Retail: $4.99 U.S.
ISBN-13: 978-0-89051-571-6

11 x 8.5 • Perfect Bound • Full Color • 88 pages
Retail: $12.99 U.S.
ISBN-13: 978-0-89051-560-0

Matter - Teacher's Guide
11 x 8.5 • Paperback • 48 pages
Retail: $4.99 U.S.
ISBN-13: 978-0-89051-561-7

Matter - Student Journal
11 x 8.5 • Paperback • 48 pages
Retail: $4.99 U.S.
ISBN-13: 978-0-89051-559-4